THE BIBLE OF THE UNIVERSE

Allan R. Rudison, Ph.D

To order additional copies of this book, contact:
Xlibris
1-888-795-4274
www.Xlibris.com
Orders@Xlibris.com

ISBN: Softcover   978-1-6698-5311-4
         Hardcover   978-1-6698-5313-8
         EBook   978-1-6698-5312-1

Print information available on the last page

Rev. date: 10/31/2022

DEDICATION

I dedicate this book to my father, Percy C. Rudison, who was the impetus that animated me for the relentless encouragement for my educational drive that I will always possess.

A C K N O W L E D G M E N T

I have the ultimate respect for all of the Astrophysicist, Astronomers, Geologist, and Mineralogist that cleared the path of the Universe for me to understand, and shed light on Dark Matter, Dark Energy, Birth of the Earth, and Life's Beginning. I look through the wide-angle lens of Isaac Newton, James Clerk Maxwell, Nicolas Copernicus, Frederick William Herschel, Michael Faraday, Galileo Galilei, Clair Cameron Patterson, Claudius Ptolemy, Edwin Hubble, Albert Einstein, William Edward Burghardt Du Bois, Carl Sagan, Robert Hazen, Stephen Hawking, Neil DeGrasse Tyson, The Carnegie Institute of Science, and NASA/JPL.

All images were taken from:

https://carnegiescience.edu/

https://www.jpl.nasa.gov/

ABOUT THE AUTHOR

Allan R. Rudison, Ph.D. is a Clinical Scientist, Consultant in Clinical Laboratory Pathology, Toxicology, and is a Molecular Biologist. His background also includes Astronomy, Astrobiology, Nuclear Medicine instrumentation, and Genetic Engineering.

He was born in Marshalltown, Iowa and currently lives in Los Angeles, California where he got his doctorate degree at the University of California at Los Angeles with the academic honors including Magna Cum Laude.

He is the President and CEO of Ethics in Science & Medicine Inc., and Nexus Global Steel & Real Estate Investments, LLC., at the Century Park Center in the Century City District of Los Angeles, California. 1(866) 923-8931

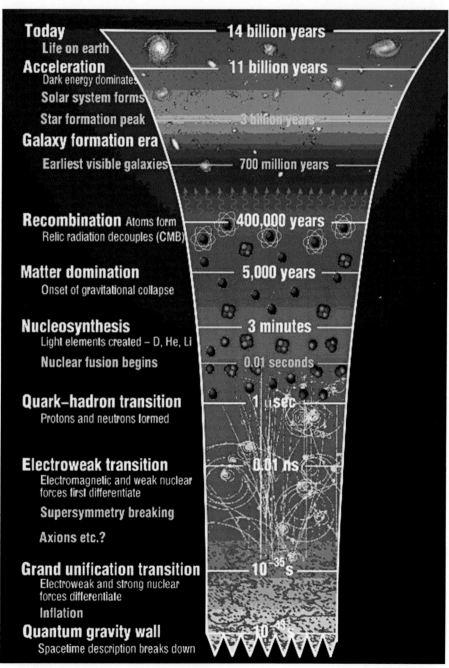

The "Big Bang" of the Universe 13.7 billion years ago

CONTENTS

CHAPTER 1

Dark Matter

Expansion of the Universe and the human brain cannot function without an electrical power supply. Prior to starting with what dark matter is, let's correlate the brains of all humans and other biological animals in this chapter to clarify the expansion concept.

The human brain is obviously capable of accomplishing inconceivable feats such as landing on the moon as well as other space flights. Some of the great current discoveries are computers and cell phones, which have significantly improved our daily lives. The Central Processing Unit (CPU), in the electronic device, is the electronic circuitry within a computer that executes instructions that make up a computer. One of the greatest discoveries of our brain includes the Central Processing Unit (CPU) for computers.

Just like all electronics, the CPU operates on an electrical power supply, as does the brain. The brain's power supply is called neurons (nerves), which carry electric charges. An example is a sinoatrial node situated on the upper part of the human heart. A signal is sent from the brain to the node that transports this electrical signal to the heart to make the heart beat/pump. Then the heart pumps blood to the rest of the body. If that sinoatrial node gets damaged, a pacemaker is implanted in its place to keep the heart beating/pumping. This pacemaker is a battery replacement. Dark matter releases that same kind of electrical energy through electrical signals to produce Dark Energy for the expansion of the universe.

Allan R. Rudison, Ph.D.

"The Universe is in Us" Human Brain Neuron Power Supply Similarities to the Dark matter Power Supply of the Universe

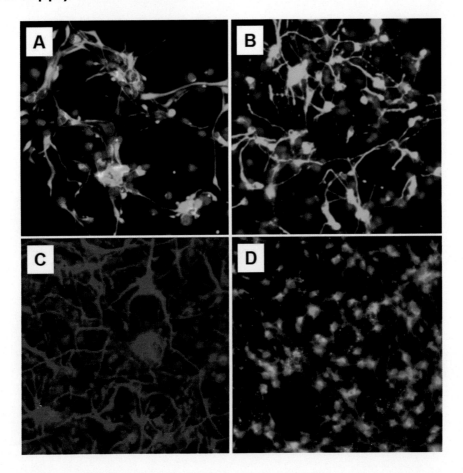

Human Neural Stem cells (HNSC) are self-renewing and are generated throughout an adult's life via neurogenesis. These multipotent adult stem cells generate the nervous system, differentiating into neurons.

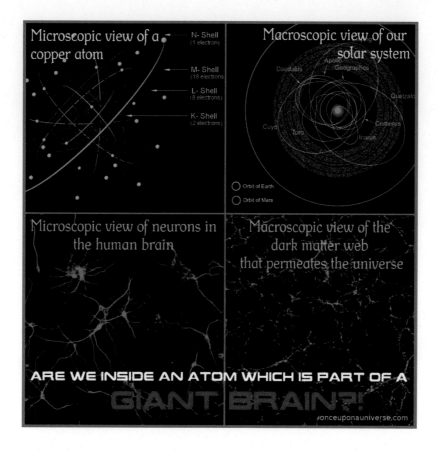

HNSC plays an important role in development, learning, and hippocampal memory. They are also used to study age-related declines, as well as neurological diseases like stroke, multiple sclerosis, and Parkinson's disease. The cells respond to injury and can be differentiated to replace lost or injured neurons. They migrate in a directed fashion to brain tumors and help replace dying neurons in injured adult brain tissue.

All human beings and other biological animals, fungi, plants, living, and non-living occupying the Earth are composed of electrical charges - both positive and negative (+ and -) just like an electronic cell. All these electrical charges emit energy. This same type of energy is referred to as Dark energy in the universe. Since Galileo Galilei in 1636, also known as the father of Observational Astronomy, invented the telescope, our exploration of the universe has expanded. With the Hubble Space Telescope in 1990, we can look deep into the universe, seeing the birth of the first galaxies in the world. It is by looking at these revelations that the biggest mystery, the cosmos was first discovered.

Dark Matter has decreased by 40% (63%-23%) from the 'Big Bang' atomic explosion of the universe. The formation of the universe after the 'Big Bang', which happened 13.7 billion years ago, Dark Matter has produced up to 72% of dark energy today. Traveling between space and time, Earth is the only place that we are sure about having the living matter of the cosmos.

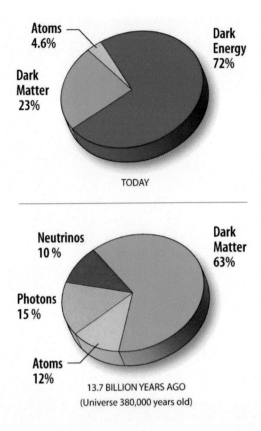

There may be an infinite number of planets across space, but our discovery of them begins from the planet we live on; Earth. We are fortunate to survive among some intelligent and passionate people, especially in the time when the quest for the unknown is at a greater prize. Humans born originally with the contents of the stars and galaxies, and now for some time inhabiting the Earth, have begun their journey. I refer to this journey as migration in the chapter to follow.

Do we have any evidence for dark matter beyond our astronomical observation? Well... yes. The neurons in our brains are our equivalent to the Dark Matter of the Universe.

How Are Humans and Dark Matter Connected?

The normal matter accounts for less than five percent of the known universe. About 23% is dark matter, and 72% is Dark Energy, both of which are invisible to the naked eye. Dark Matter in the universe is directly proportional to the dark matter (neurons) in our brains. This is why "The Universe in Us". What follows dark matter is dark energy. We will further discuss this in the upcoming chapters.

In 1929, Edward Hubble examined how the wavelength of light emitted by distant galaxies shifts towards the red end of the electromagnetic spectrum as it travels through space. He found that the fainter, more distant galaxies showed a large degree of redshift. He determined this was because the Universe itself is expanding.

The redshift occurs because the wavelengths of light are stretched as the Universe expands. Recent discoveries have shown that the expansion of the universe is accelerating. Before that, it was thought that the pull of gravity would cause the expansion to either slow down or even retract and collapse in on itself at some point.

Space does not change its properties as it expands. New space is constantly created everywhere, as in our brains, where neurons (nerve cells) continue to multiply to be innovative. Galaxies are tight clusters held together by gravity, which is why we do not experience this expansion in our daily lives, but we see it everywhere around us. Wherever there is empty space in the Universe, more is forming every second, similar to our brain development.

Therefore, Dark Energy seems to be some kind of energy, intrinsic to empty space. This energy is stronger than anything else we know, and it keeps getting stronger as time passes by. Empty space has more energy than everything else in the universe combined. Dark energy is a property of space. Empty space has its own energy. It can generate more space and is quite active.

As the universe expands, so does our brains – allowing us to come up with the technology of cell phones, computers, which get updates, improving with time from gigabytes, megabytes, to terabytes in manual devices. This is the reason for updates on devices like the cell-phones-iOS (internet operating system), computers-CPU (Central Processing Units), and cars-GPS (Global Positioning System). These are the examples of the expanding Universe correlation with our expanding brain that makes these devices for our daily needs of survival. No matter how much we feel we are on top of things, we are still very much occupied with smart-phones and computers developed by the human brain to assist us as we continue to evolve.

CHAPTER 2

Dark Energy

The properties that affect the expansion of the Universe are twofold. One of them is that the vacuum of space is persistent. As the Universe expands, there is more space, so there is more vacuum and more space for energy produced by Dark Matter. Consequently, in our evolving Universe, we find more Dark Energy.

Our best evidence is that there is something invisible all around us. Even though we have never seen it, like gravity, it makes up 72% of the Universe. What is it called? It is called Dark Energy. Twenty years ago, looking at supernovae, astronomers discovered that the Universe was not just expanding outward, it was actually accelerating faster and faster outward. By this, scientists began to think, maybe space itself was expanding.

In Einstein's view, space-time is basically a mesh that just sits there unless energy or matter bends it. You might have seen videos or images of the bending of that space-time mesh which causes things to attract. Maybe, there is a special kind of material, something invisible and undetectable that pushes the space-time to mesh apart. Imagine that space is full of particles that repel each other. Whenever there is enough space for a new particle to form, it does form. That particle repels all the other particles, and then, more particles pop into existence, pushing each other apart, and pushing space with it.

Space and time are not made of electronic particles, but on a really small scale, you actually have particles popping in and out of existence. They are called virtual particles, and they live for a short amount of time before they annihilate each other. The energy of these particles causes the Universe to expand. If you calculate the amount of push from all the virtual particles that exist in the standard model of particle physics, you will get too much pushing. You actually get a trillion and zillion times pushing. Our current quantum theories of particle physics cannot explain observations of Dark Energy.

Space itself expands without worrying about these particles. Dark Energy is even hard to imagine. It has the same density everywhere. If it starts to swell, it starts expanding gradually. It does not even get diluted because it is embedded in space. As space expands, you get more space and more Dark Energy, which all looks the same.

For now, all we know is that, the bigger the Universe gets, the faster it falls apart. One day, the distance between our milky way and the neighboring galaxies will grow faster than the speed of light. When that happens, even our best telescopes that exist today, may not be able to see those galaxies. The second vacuum property, the key to expansion is that the vacuum of the Universe also has pressure.

This pressure repels the expansion of the Universe as it tries to pull on the expanding Universe like the tension between the Sun and the Moon, which causes the waves in the oceans. It is a lot similar to a tug of war between these two objects, because they both are huge rocky metal magnets, continuously rotating. This constant movement between these two objects creates waves in the ocean, constantly moving.

I am not going to delve into the equations of general relativity here because that is not the purpose of this book. The purpose of this book is to inform its readers about their brain's connection to the expansion of the Universe, the fact that "The Universe in us", and the behavior of the expansion of the Universe relative to the expansion of our brains.

The definitive irony is that even though there is more and more Dark Energy in the Universe, its own opposition to that expansion causes the whole equation to hasten the growth of the Universe. That is because the power supply of Dark Matter is much more influential than the resistance of the vacuum of the Universe.

To further simplify my concept of Dark Energy, I want to refer you to the Chinese belief: Yin and Yang, it defines how opposite or contrary forces are actually complementary, interrelated, and symbiotic in the natural world, and how they give rise to each other as they interconnect.

Dark Energy and Gravity are like Yin and Yang of the Universe. The expansion of the Universe is controlled by both the forces of gravity, which act to slow it down and Dark Energy, which pushes matter and space apart. In fact, dark energy is pushing the cosmos apart at a faster rate, causing the Universe's expansion to accelerate.

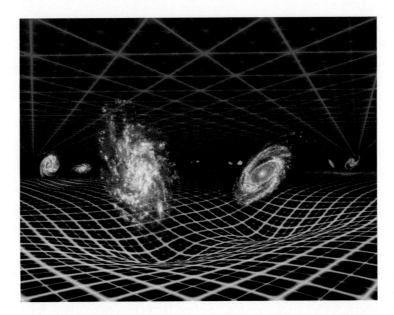

Dark Energy, in the figure above, is the top grid. Gravity is the green grid that bends as it is known to do. The expansion of the Universe is controlled by the force of gravity, which slows down expansion. Dark energy is so powerful that it forces the Universe to very high speeds causing the expansion to accelerate. Dark energy presently occupies 72% of the Universe and dark matter 23%.

As stated in Chapter 1, Dark matter was 63% at the beginning of the Universe, 13.7 billion years ago, along with 37% of atoms, photons, and neutrons. Thus, electrical particles complete 100% of the matter in the vacuum of space called the Universe. At this modern rate of expansion, the Universe will run out of Dark Energy and will eventually stop expanding. Humans will then have to rely on the knowledge they obtained from the expansion of the Universe and brain expansion for survival, somewhere other than the earth, because earth may no longer exist at that time.

This is why humans are currently exploring other planets, mainly the Moon and Mars, our closest neighbors for the survival of our species, as well as other living species that occupy the earth that we rely on for our balance of nature. There will be more discussion of our survival in the chapter with the discussion of evolution.

Dark Energy/Hadron Collider

Proof of Dark Energy comes from various sources. Cosmologists have long believed that Universe, as a whole, has a unique shape. It can be closed or confusing to the surface of a sphere, open similar to the surface of a saddle, or flat. The precise shape that the Universe evolves into depends on its total mass-energy composition. Astronomy has measured the shape of the Universe by observing the cosmic microwave background radiation that occupies all of the space nearly uniformly and is a remnant of the Big Bang.

The shape of the Universe has an impact on the size of the slight variations that we see in the cosmic microwave background. The evaluation of these deviations tell us that the Universe is indeed flat. Interestingly, we are sure from observing the galaxies in clusters that all matter in the Universe is comprised of 23 percent of the total mass-energy density needed for a flat Universe.

In order for our observations of clusters to match with those of the cosmic microwave background, we need to record the missing 72 percent of Dark Energy. In addition, evidence for more direct energy comes from supernova observation outburst that occurs when a gigantic star reaches the end of its life, and for a short period, they shine as brightly as an entire galaxy of 100 billion stars.

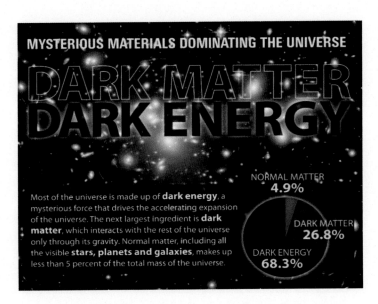

Astronomers have used these outbursts as *standard candles* because the ignition we receive from a supernova always follows the same pattern with time. We can hence deduce how bright the supernova really is and how bright it looks on the sky. The uniqueness of dark energy is no longer a mystery to me, but the modest explanation is known as the cosmological constant because its density is persistent over both space and time.

It is also known as vacuum energy, as it is thought to be the inherent energy of empty space. In many ways, the sophisticated solution to the problems of Dark Energy, because it defines many cosmological observations with one simple figure; the density of the vacuum energy is a cosmological constant.

Large Hadron colliders in Cern

Dark Energy and Dark Matter together make up 95% of the Universe. This only leaves a small 5% for all the matter and energy we are familiar with. This book explains what Dark Matter and Dark Energy really are relative to the expansion of the Universe, and the correlation of the expansion of the human brain. We have known for about twenty years that Universe is accelerating every day, and our cosmos is expanding with an increasing rate and multiple independent lines of evidence.

This all points to the same conclusion; the accelerating expansion of the Universe is occurring at the same rate as the expansion of the human brain.

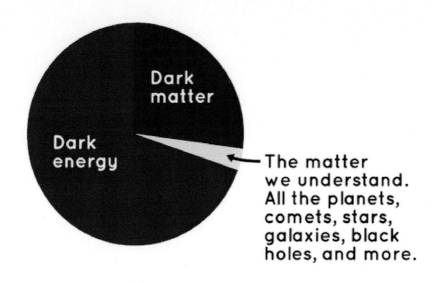

This is and has been for as long as the brain has developed is referred to as *evolution* ever since Charles Darwin's era; the 1800s. Evolution is the basis for "The Universe is in Us", including Dark Matter, Dark Energy in concert with the accelerating expansion of the brain.

Fortunately, our brain has reproduced itself in addition to its nomenclature to be innovative enough to provide us with computers, cell phones, and GPS systems through our evolution to assist the brain with creating devices that make our lives better, but also cause problems, i.e. hackers. However, this is also part of innovation. We have to manage and solve these issues. There are positive and negative aspects in all walks of life, just like the positive and negative electrical charges that are directly proportional to life.

Large Hadron Collider

The Large Hadron Collider could reveal our origin. The biggest scientific machine ever built has begun churning out the smallest known bits of matter in the Universe. Its goal is to uncover some of the deepest, long-hidden secrets of nature. This enormously ambitious device is the Large Hadron Collider, a 17-mile-long ring-shaped tunnel 300 feet under the Swiss-French border near Geneva. Its development began in 1954, and it has been hailed as the greatest scientific undertaking since the Manhattan Project, which created the atomic bomb during World War II. Some scientists regard the Hadron Collider as nothing less than a 'time machine' that may let them go back billions of years to study the origins of the Universe. They hope it will shed light on profound questions such as: What happened immediately after the birth of the Universe 13.7 billion years ago? What does the Universe consist of?

How did matter come to be? Are there more dimensions of space than the three (plus time) that we're familiar with? The aim is to recreate the conditions that existed in the initial moments after the Big Bang, the beginning of the Universe 13.7 billion years ago, as discussed in Chapter 1.

The answers to those questions may have no immediate practical applications, but history has shown repeatedly that advances in fundamental science usually lead to useful things such as telephones, radios, computers, improved manufacturing, nuclear energy, global positioning satellites, and so on.

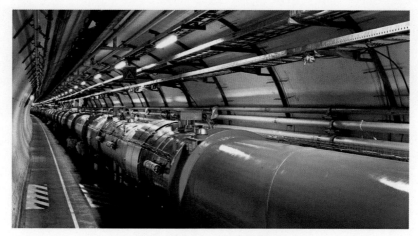

Large Hadron Collider

After World War II, science and technology have been accountable for half of America's economic progress. The Hadron Collider is another large innovation because it is the biggest collection of scientific tools ever gathered in one place. It works with hadrons physicist jargons, protons, and neutrons (electrical particles) that make up the nucleus of an atom.

It is a collider that smashes protons, tiny subatomic particles so that scientists can peer at their crushed innards. The secretive ring is lined with 1,232 50-foot-long magnets, each weighing 35 tons. The tunnel contains four huge particle detectors, as big as apartment buildings, plus two smaller ones, all crowded with scientific instruments. The whole complex is chilled by 120 tons of liquid helium to almost absolute zero, even colder than outer space. The builder and operator are the European Organization for Nuclear Research, commonly known as CERN, an acronym for its original name that translates a bit differently as it is in French. CERN began as a partnership of 12 nations in 1952 and now has 20 members.

The U.S. joined it as an observer, not a full member, in 1997. Hundreds of Americans currently work here, along with a rotating cast of more than 10,000 scientists, engineers, and technicians from around the world.

The Hadron Collider functions as follows. Bunches of protons, the nuclei of hydrogen atoms, are injected into the tunnel by a series of accelerators that boost their speed to 99.9% of the speed of light.

The speed of light is 671 million miles/hr. The magnets direct the protons around the ring 11,245 times a second in rigorous beams.

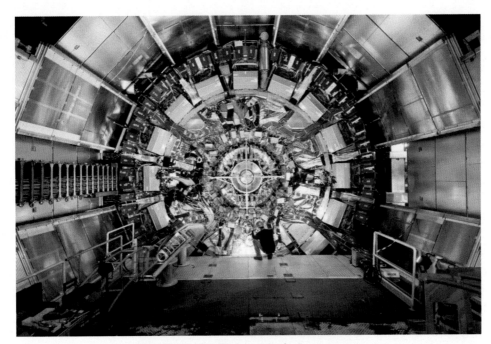

Hadron Collider Maintenance

They zip through the tunnel in two parallel pipes, half of them moving clockwise, half counter-clockwise. At four locations, the beams cross each other, colliding protons into each other head-on. After the protons collide, they produce a spray of even tinier particles, mostly *quarks*, which are the smallest, most essential building blocks of matter so far discovered. As per CERN, protons are 100,000 times smaller than the simplest atom of hydrogen. Quarks are 10,000 times smaller than protons. For contrast, if a hydrogen atom is six miles across, a quark will still measure less than four-thousandths of an inch. These collisions are extremely powerful. Physicists measure a collider's energy in electron volts, which is the tiny force required to move one electron from one side of a one-volt battery to the other side.

The Hadron Collider produces seven trillion electron volts, abbreviated as 7 TeV of energy. That sounds like a lot, but protons are so small that a little energy goes a long way. To a human, one trillion electron volts feel like a mosquito biting your skin. The Hadron Collider is the most powerful proton collider in the world. The current record is 2 TeV at the Fermi National Accelerator Laboratory in Batavia, ILL.

In a few years, if all goes well, the Hadron Collider's power will be doubled to 14 TeV. What we hope to learn by slamming subatomic particles together? At that time, the infant Universe was so hot that it consisted of only a hot soup of quarks, and another breed of particles called gluons. Unlike quarks, gluons have no mass and are not constituents of matter, but they carry the force that holds quarks together. After a few small seconds, the primeval quark-gluon soup, or plasma, cools enough to allow quarks and gluons to form protons and neutrons, which eventually combine with electrons (+-) charges to make atoms, molecules, stars, and people.

This is the evolution of the power supply of dark matter, converting to Dark Energy expanding the Universe. One of the major goals of the project is to explain how quarks combine to create matter. Scientists hope they will find in the debris of the collisions, evidence of undiscovered subatomic particles called the *Higgs boson*, named after a Scottish physicist who predicted such a particle in 1964.

Like the gluon, bosons are particles that have no mass but carry a force. Scientists think the Higgs boson if it exists, is the particle that allows energy to turn into mass. The theory is that Higgs bosons are spread throughout the Universe, like flowers in a field. Particles acquire mass. In other words, it becomes matter. This matter is sometimes referred to as the entire human being. CERN estimates the collider produces about 700 megabytes of data per second, enough to fill a 12-mile-high stack of CDs per year. The material will be distributed worldwide over the Internet. The cost of the Hadron Collider is about $5 billion so far for design, construction, and operation. The proton collisions yield tiny *fireballs* thousands of times hotter than the sun, but they last only microseconds. Scientists say they are not hazardous.

According to CERN, the fireballs' energy is much weaker than the cosmic rays from outer space that have barraged earth harmlessly for billions of years. However, for being in the medical field for many years, I do not think so. We strongly believe that the bombarding of cosmic rays causes serious cancer risks, especially skin cancer. Tours are available for the Hadron Collider on YouTube.

Hadron Collider Maintenance

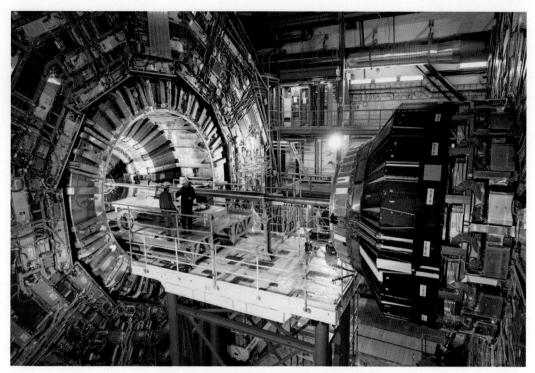

Hadron Collider Maintenance

Allan R. Rudison, Ph.D.

Hadron Collider Maintenance

CHAPTER 3
The Universe/Brain/Internet Connection

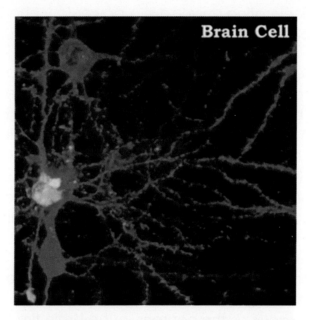

Here A Brain Cell Is Compared To The Filamentous Arrangement Of Galactic Universe Superclusters

This photograph shows a comparison between the human brain cell and the Universe. Remarkably, physicists have found proof that the Universe acts as a giant brain. Moreover, a human brain grows in very interesting ways, by connecting clusters of brain cells to each other using neurons as pathways.

The electrical firing and linking to these pockets keep expanding and connecting to form a massive complex network, very much like the internet. The original study puts it this way; we show that the causal network representing the large-scale structure of space-time in our accelerating Universe is a power-law chart with strong grouping, similar to many complex networks such as the Internet, social, or biological networks.

We proved this structural similarity is a consequence of the asymptotic equivalence between the large-scale growth dynamics of complex networks, and causal networks. Science has always been suggesting that consciousness might be the underlying fabric of the Universe, and that it spawns everything. The Giant Brain theory is even more compelling now that we know our Universe's fabric grows like a brain in addition to being conscious. There is a reason that the Universe grows just like a brain, or the brain grows just like the universe simply because "The Universe is in Us". Scientists often speak of the Universe being a reflection of ourselves, and point to how the eye, veins, and brain cells mirror our visual phenomenon in the natural Universe.

It's because the Universe is a giant brain. The fact is that it has been proven by scientists that the Universe acts as a giant brain with the electrical firing that has shaped the expanding galaxies exactly as the neurons in the brain fire electrical impulses from the brain to other organs of the body for thought and motion.

The results of a computer simulation suggest that natural growth dynamics, the way the systems evolve, are the same for different kinds of networks whether it is the internet, the human brain, or the Universe. When the data was collected from the scientists for this project of the history of the Universe with evolutionary growth of social networks and brain circuits, the finding was that all the networks expanded in similar ways in that they balanced links between similar nodes with ones that already had connections. Someone surfing the internet usually will use Google, Bing, or perhaps Yahoo and may visit YouTube. Brain cells connect to other brain cells and link to other cells that are hooked up to loads of other brain cells. In the brain, there is information coming in called Sensory Information and information going out called Motor information. This happens at a very high speed in conversation and learning.

The Universe has a built-in mechanism that has formed to expand and evolve in a way that encourages novelty, and an increase incoherence. It is the Dark Matter power supply resulting in Dark Energy for the expansion of the Universe. Our minds are microcosms of what is happening within the expansion of galaxies and celestial bodies since our minds are activities of the Universe, and the laws that govern it.

We are interconnected to the Universe and its evolutionary process more than I had previously realized. Now that the Dark Energy of the Universe produced from the Dark Matter expanded enough to correlate that same expansion of our brains, I can now shed that light of the Universe/Brain connection with my audience via this book. That same correlation is the reason for our computers, cell phones, GPS, etc. We are witnessing the Universe expansion of Dark Energy that is a direct correlation with our Brain expansion as an evolutionary process. We are on the Earth, and the Earth is in the Universe. "The Universe in Us". When we look up at the stars and their formations, we are looking at the same intelligent process that is responsible for the eye creation we are looking through, and the neural networks that are interpreting the data.

The next photos shown are of the causal network representing the large-scale structure of space-time in our accelerating Universe is a power-law graph with strong clustering, similar to many complex networks such as the internet, social, or biological networks. All of these networks in the Universe, and on Earth, are intertwined as you will witness in the following chapters.

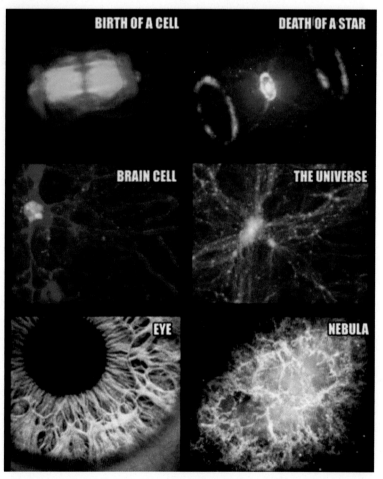

Human anatomy has evolved from the Universe

In the chapters to follow, the earth's development will clarify the factors in the evolution of life that the Universe plays the most significant role providing the minerals, gases, and electricity, that are necessary for the development of the cycle of life.

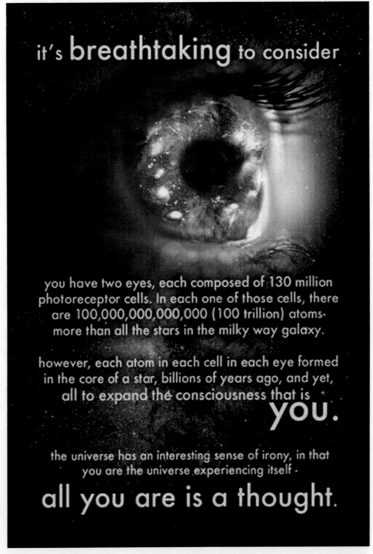

Our brains are a thinking machine of thoughts!

The evolution of Dark Matter of the Universe to Dark Energy 1billion-14 billion years is similar to the neuron development of the human brain evolution, but of course, the evolutionary process of the brain was accomplished in a much shorter time span of approximately 3 million years ago.

1 Billion yrs.- 4 Billion yrs. -14 Billion yrs

Science is founded on fact. The fact is that the Universe is simply a great machine that came into being as the result of the 'Big Bang', which came from the clash of atomic particles, as demonstrated by the Hadron Collider. The human being is no exception to the natural order. Humans, like the Universe, are a machine.

Nothing enters our minds, or determines our actions, which is not directly or indirectly a response to stimuli. A single fundamental law of nature governs these networks, which is a signal that some misunderstanding of how nature works were in question until now. There is now scientific evidence that there is a direct correlation between the expansion of the Universe and the expansion of the brain.

We recognize it as another chapter of evolution. In 1968 we began our journey home to the Universe with the landing on the moon. As the Universe continues to accelerate its neuron-like expansion, it is in direct correlation with the brain expansion of neurons. There are always hiccups, and bumps in the road of this evolutionary process. There's the Ice Age, Slavery, Global warming (Climate Change), Wars, and Assaults on Government Institutions, but there is no way that human beings can stop the persistent, unstoppable evolutionary process of the Universe, and the evolution of Mankind.

The Universe began with Physics through the explosion of Atoms, (+ -) electrical particles (protons, neutrons, and electrons) clashing together, which causes explosions as demonstrated by the Hadron Collider, but with much greater force than the Atomic Bomb. Then later came the formation of planets. Then Chemistry with the presence of water as the planet cooled, and the oxygen formation from the oceans, then Biology, with the formation of life. The same principles apply today with the evolution of the science of technology with our computer and cell phone devices to assist us with everyday life challenges as the brain continues to develop with the evolutionary process.

The CPU is a Computer Processing Unit. It functions as an electronic circuit within a computer for basic operations. The Brains neurons are the circuitry within the Brain for the same function. Dark Matter is the Power supply within the Universe. RAM-Random

Access Memory is Memory Storage. The Hippocampus located in the brain is the memory storage of the brain. The Router is a computerized device that forwards data from one network to another. The Thalamus in the brain is referred to as "The Router" and also "The Grand Central Station" because it sorts data coming into the brain and sends the data where it needs to go.

Electricity must have a system of protection such as rubberized cords from your electrical outlets to your microwave, and for charging your computer or your cell phones. The spinal cord has a myelin Sheath to protect your neuron signals from your brain throughout the rest of your body. Dark Matter has the protection of electromagnetic radiation to send wireless electrical currents to create Dark Energy for the expansion of the Universe. X-rays and computers function by electromagnetic radiation.

This is why we have wireless connections today with computers and MRI functions in hospitals for imaging. Nearly all hospital instrumentation for scanning body parts are developed from the research of the Universe functions such as electromagnetic radiation for X-Rays, and scanning techniques of the brain and body parts for diagnosis of cancer and other life-threatening organ complications.

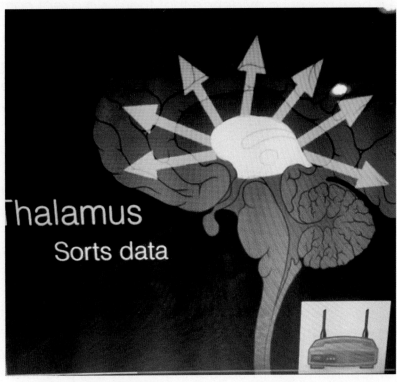

Thalamus Router   **Computer Router**
Sorts Data   **Sorts Data**

Mathematical methods of algebra have helped find structures, and multidimensional geometric spaces in brain networks. The human brain is home to structures and shapes that have up to 11 dimensions. Human brains are estimated to be home to a staggering 86-100 million neurons, with mathematical methods of algebra helping us find structures, and multidimensional geometric spaces in brain networks. The human brain is home to structures and shapes that have up to 11 dimensions. Human brains have several connections from each cell webbing in every direction possible, forming a vast network that makes us capable of thought, and consciousness.

With approximately 86-100 billion neurons in the adult human brain, each neuron makes approximately 1,000 to 10,000 contacts with other neurons. It has been calculated by scientists that the number of combinations of brain activity exceeds the number of elementary particles in the Universe. There's a multidimensional Universe inside our brains.

A Multidimensional Universe Inside Our Brain

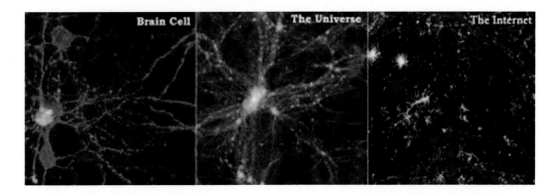

These photos show neurons in the brain, which show striking similarities to the network formations in the Universe, and the internet because they are interrelated. This network is the basis of our connection to the Universe. "The Universe is in Us". The neurons' main job is to generate an electrical signal called an Action Potential, which it does act if sufficiently excited by other neurons.

The Action Potential of a single neuron can then generate their own signals that travel to and stimulate other neurons to which they are connected, creating a network of neurons that perform a specific brain function.

This action is the same action generated from your computers CPU, and Dark Matter to Dark Energy for expansion of the Universe. To set similarities aside, it is important to remember that the difference in scale is enormous. Galactic Clusters are millions and billions of light-years across the Universe. If that is put into perspective, light travels at a speed of 671 million miles per hour/186,282 miles per second. Light can travel around the world/Earth seven (7) times in one second, but it takes light millions of years to travel from one side of a super galactic cluster to another.

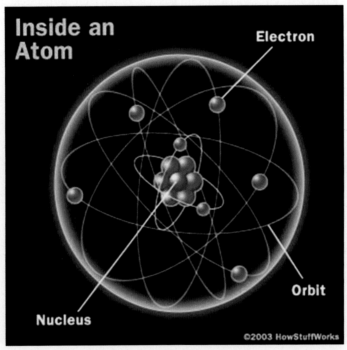

Electrons Orbit a Nucleus

The unfathomable enormity of the Universe forces us to revisit our place in it constantly. On a Universal scale, we are microscopic. It's as if we are on an electron orbiting nucleus in an atom. Your brain uses neurotransmitters to tell your heart to beat/pump, your lungs to breath, and your stomach to digest. They can also affect mood, sleep, concentration, weight, and other problems when they are out of balance. Neurotransmitter levels can be depleted in many ways by stress, poor diet, genetic predisposition, and recreational and prescription drugs. Alcohol, and caffeine use can cause their levels to be out of range also.

When your neurotransmitters are balanced, you can think more clearly, feel happier, and experience well-being more often. You can also handle stressful situations more effectively. To balance the neurotransmitters in your brain, you can maintain a good diet, and avoid alcohol and caffeine on a regular basis.

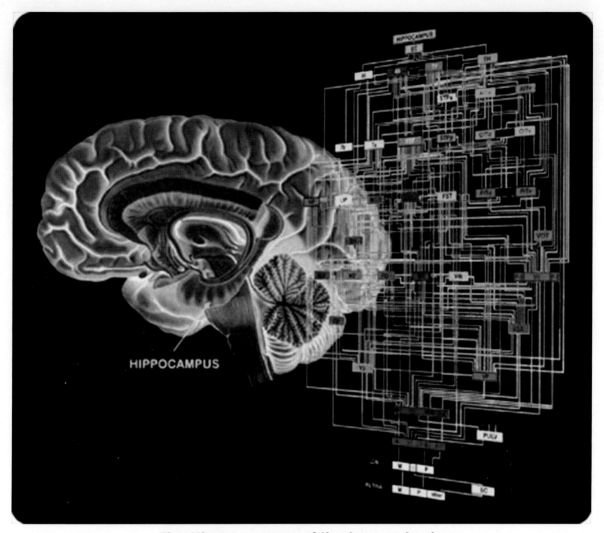

The Hippocampus of the human brain

The Hippocampus functions as the memory and storage section of the brain, as does the (RAM) Random Access Memory section of the computer. The power supply of the brain is neurons with positive and negative (+ -) electric charges causing synapses to fire, which describes nature at the smallest scales of energy levels of atoms and subatomic particles, which make up the entire Universe.

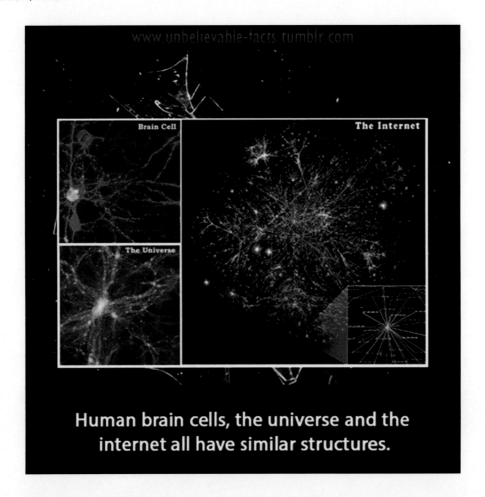

Human brain cells, the universe and the internet all have similar structures.

All three structures have branches extending from the neurons in the brain from the brain cells, the branches extending from the Galactic Electric Super-Cluster reactions in the Universe, and branches extending from electrical responses of the Internet.

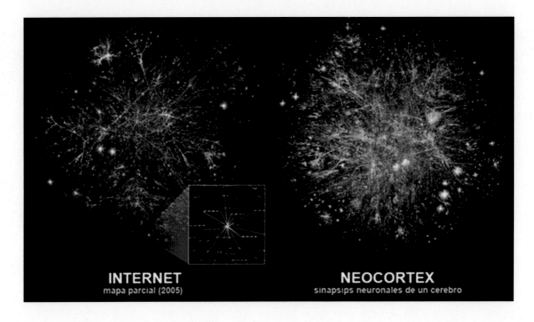

INTERNET
mapa parcial (2005)

NEOCORTEX
sinapsips neuronales de un cerebro

The Internet, and the Neocortex, 'the frontal lobe section of the brain' shows striking similarities because of the "Universe Brain Connection", and the creation of the internet by the human brain as an assistant for the human brain in concert with the acceleration, and development of technology that we experience every today. The expansion of the Universe from the Dark Matter power supply transformed to Dark Energy, is the correlation of evolution of the Universe to the development of life on the planet earth.

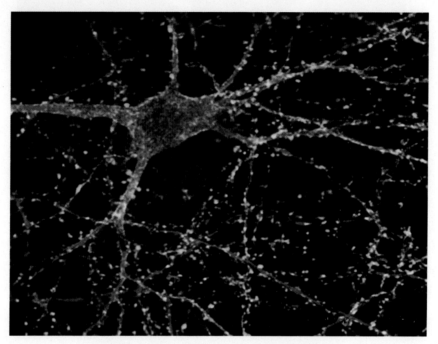

Brain cells with nerve transmitters

The giant neuron (nerve cell) is connected by branches called Axons and Dendrites, that connect to other parts of the brain, and transmits electrical signals of information. The same sequence happens from Dark Matter electrical signals to create Dark Energy for expansion. The same response occurs when a computer transmits signals from the CPU to the router, and the router sends data to other parts of the computer system like your printer. They all operate the same way because we are all connected. Neurons fire through branches of Axons and Dendrite Neural transmitters

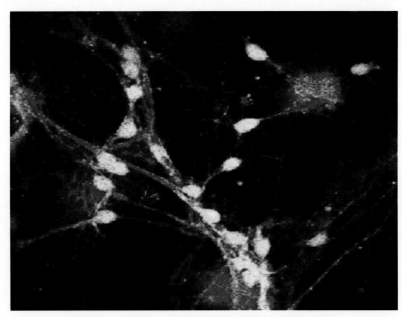

The synapse firing of brain neurons

The firing of electrical neurons transmitting signals to and from the brain is responsible for sensory signals of information coming into the brain, and motor signals going out of the brain for thinking, processing information, and communication. These electrical firings of the brain begin at birth, and extend throughout life until death when all electricity shuts down, which is the same reaction when you shut down your computer.

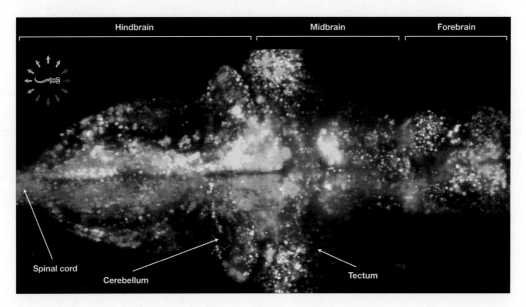

This is a photo of a Zebrafish neurons firing all through its Hindbrain, Midbrain, and Forebrain as it is functioning during its searching mode. In a tenth of a second after a neuron fires, the indicator becomes fluorescent, and glows. This is all an indicator of electrical reactions that are the same as Dark Matter electrical reactions from its power supply, creating energy to expand the Universe via Dark Energy. Also, the electrical responses are the source of expansion of our brain for evolution and innovation, such as space launching, computers, cell phones, and energy for cars run by electricity.

When humans take some time to consider all of the matter they are made of, it's a monumental task to discover all of the mechanisms in which they are constructed. I realize how mind-boggling this can be, and this is why the author of this book is bringing it to your attention, so you can appreciate all of the evolution that has transpired to make you what you are. We are a mass of matter created by the Universe "Big Bang" explosion run by a power supply of electricity.

CHAPTER 4

The Birth of the Earth

The previous chapters have basically laid the foundation of our origin from the birth of the Universe 13.7 billion years ago, to the 'Birth of the Earth' 4.5 billion years ago. All of the constituents of life on earth originated in the Universe, and are the basis of all life present on earth. The formation of the earth strongly suggests that the earth was formed from the clashes of small rocks, larger rocks, and dust from more encounters that became meteorites in the Universe, then further collisions that became planets to create what we call the Earth.

Now we observe the earth as volcanic rock, mountains, and water produced from different stages of heat and cooling of the Earth that created beautiful views rising from land, and oceans. There are many intriguing sites in the world that are remnants of the collisions of meteorites marked by heating and cooling of the primitive earth to form such sites as Niagara Falls, volcano's, and many historic national parks. The collisions of these rocks and dust from the 'Big Bang' origin of the Universe played a significant role in the recipe for life, because of their mineral constituents such as calcium carbonate. Calcium Carbonate is the most responsible constituent for the making of teeth, and bones for many creatures of life. We have an absolute dependence on rocks, and its components for so many things such as buildings, homes, streets, satellites, computers, cell phones, jewelry, wedding rings, to name a few important ones that are in use around the world by most people on a daily basis. Not only is "The Universe in Us", it's all around us.

When there was a collision of massive Meteorites during the development of the planets, smaller rocks, and dust scattered all over the Universe, creating massive high temperatures of heat, and dust developing formations of other structures as planets. This is how the "Birth of the Earth" began. There will be some photo illustrations following this page to unpack some of the stages that the earth went through for its formation to the development of the Earth as we view it currently. Giant meteorites of iron that circled the sun clashed, and began the creation of all planets in the Universe.

It was the clashing of these meteorites that built all planets in the Universe. Our planet was created out of rocky mineral-filled iron meteorites, and dust from the Big Bang explosion 13.7 billion years ago. In the following photo, meteorites are about to crash into each other, and explode into a fiery and molten rock formation that later becomes rocky volcanic mountains. The meteorites are traveling at thousands of miles per hour, and the explosions are like atomic bombs. Inside the giant rocks are the same minerals, iron, and other chemical ingredients that are used today for our daily necessities, and recipes of life.

Meteorites in motion near-collision forming earth

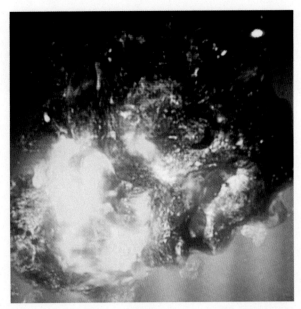

Volatile meteorite collisions developing planets

Planets are giant magnets of iron ore located in their inner structures. This is why they circle the Sun bound by the magnetic force they have on each other. It's the same concept for the waves in the ocean from the magnetic force pulling on the earth by the moon, and vice versa for the circling of the Earth, and the Moon and other planets around the sun.

After the clashes of the large meteorites circling the Sun, the planet was black with smoke and fire. There was extreme heat that caused the early earth to be molten with temperatures of thousands of degrees, causing the smoke and blackness of the planet Earth in its early stages. Because the planet was in a cold Universe vacuum, the planet would start to cool, and temperatures would decrease over time while circling the sun. The circling of these forming planets orbit the sun because they are giant magnets of iron pulling on each other as does the earth and the moon, causing waves in the ocean.

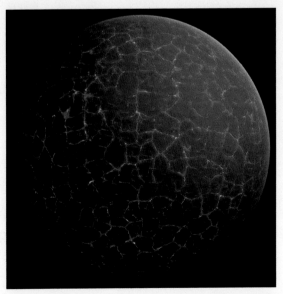

Early hot molten Earth burning after collisions

This hot molten earth was the beginning of the formation of volcano's that would later be spewing out hot lava from deep inside the center of the planet. In that hot lava would be the beginning of minerals such as crystals and diamonds, and other precious minerals that were contained in the meteorites that are constituents of life on earth.

Stages of burning of Earth in its development

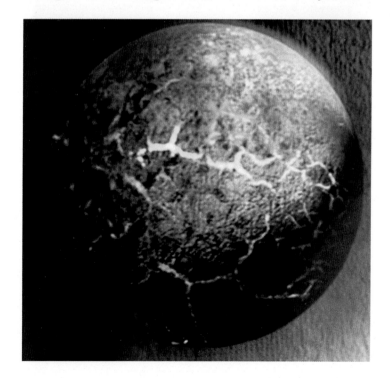

Smoke and Molten Black Planet forming Earth

Volcano's began to spew hot Lava inside the planet. When it cooled, it covered the earth with its first formation of black rock with a lot of smoke. Some of the mineral ingredients that life would later need were outside, and inside these lava spewed rocks.

Earth cooled to grey in a cold Universe

Deep in the ocean, approximately one mile on the ocean floor, underwater volcanic mineral-rich hydrothermal vents with temperatures of 600 degrees were found and considered to be the center for the origin of life. There were thriving living microbes from chemical energy from the hydrothermal vents. Only bacterial growth could have withstood those kinds of temperatures.

Volcano's began to spew hot lava iron from the core

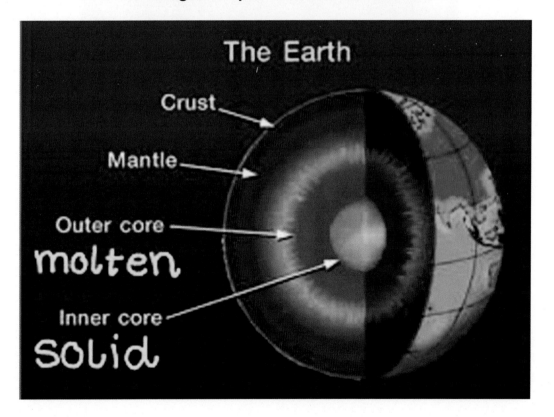

Hot lava rock and minerals from volcano's

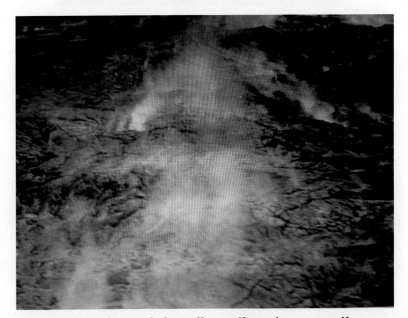

Volcanic rock for all continents on earth

The gases coming out of the volcanic ash combined with the other gases in the atmosphere set the stage for habitable conditions for life. The illustrations below represent those conditions along with the minerals and crystals inside volcanic rocks.

Colorful rocks originating from meteorites

Polished volcanic rocks full of minerals and chemicals

Allan R. Rudison, Ph.D.

The colorful minerals inside these rocks are quartz and diamonds. The mineral composition inside the rocks has chemical, and physical properties that are crystals. These elements inside the rocks are natures building blocks of societies of life. That's why rocks are essential in our modern life to make large buildings like skyscrapers bigger, and taller, computers, televisions, automobiles, and cell phones. These mineral elements of lava rock added to steel make steel stronger, or added the mineral cobalt to cell phones, and they make the batteries last longer. The mineral properties were also essential in creating life. Earth began during the collisions with other meteorites circling the sun, and additional minerals were created in the intense heat of the collisions and development of earth's planet. The development of the extra minerals changed the appearance to grey.

The grey color of the earth came from the granite that was formed with minerals and chemicals from the clashing meteorites in the Universe, like the photos of Yosemite National Park. The rocky mountain formations set the stage for all the planets' continents.

Yosemite National Park cliffs are granite containing quarts, and other minerals. Granite from the meteorites are the foundation rocks of all continents, and the granite is responsible for the grey period of the color change of the Earth's development. The next stage, the earth went through is the water stage, that turned the earth from grey to blue.

Yosemite Park Cliffs of Granite from meteorites

Mountain formations from clashing meteorites

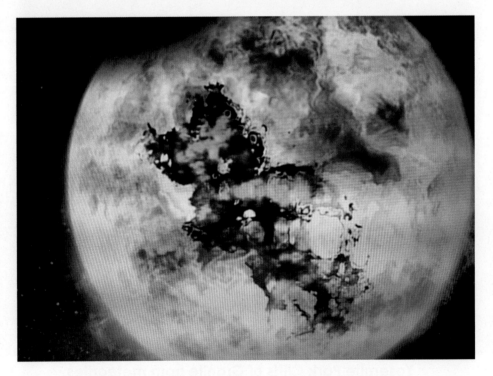

The earth began to turn grey from granite.

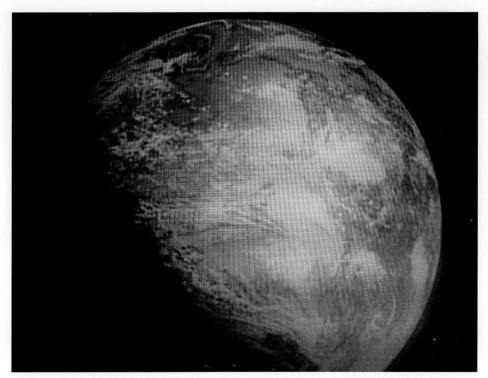

The earth started to turn white in a cold Universe

The earth began to melt circling the Sun

The water began to form all the earth's ocean

Zircons are the oldest microscopic mineral crystals on earth, showing the environment and age when the zircon crystals were formed nearly 100 million years after the formation of the earth, 4.5 billion years ago. They are the oldest particles of the earth ever discovered, shedding light on the earth's appearance at that time. The zircon crystal is, in fact, a "time machine", validating that the earth was still covered with Lava. The zircon was estimated to be 4.3 billion years old could only be formed in the presence of water. Since water is a key element of life, the zircon suggests that the earth was a habitable environment for life of some kind, most likely bacterial life. The probability of life began 4.3 billion years ago. Laboratory experiments proof of early earth's environmental atmosphere, naturally containing gases, electricity, water molecules, and when minerals from rocks were added to the mixture, that mixture produced amino acids, which are the building blocks of life creating protein for muscles and body tissues.

The earth began to warm, and turn blue as it started to thaw while circling the sun. This was the first sign of water on the blue warmer planet. The water plus the minerals from meteorites is the essential ingredient in the origin of life. Water dissolves all the necessary molecular ingredients of life, and combines them in their interactions. The initial point of life is water. The earth cooled enough to have water for oceans based on zircon crystals, showing the environment and age when the zircon crystals were formed nearly 100 million years after the formation of the earth, 4.5 billion years ago. They are the oldest particles of the earth ever discovered, shedding light on the earth's appearance at that time.

The zircon crystal is, in fact, a "time machine" validating that the earth was still covered with Lava. The zircon was estimated to be 4.3 billion years old could only be formed in the presence of water. Since water is a key element of life, the zircon suggests that the earth was a habitable environment for life of some kind, most likely bacterial life. The probability of life began 4.3 billion years ago.

In 1871 Charles Darwin suggested that life may have begun in a heated shallow pond with all the amino acid complex building blocks of life getting energy from the sun, still he didn't have access to the ocean floor, nor did he know that to have amino acid production, electricity had to be the kick-starter for amino acid production. In 1895, life was first discovered on the ocean floor nearly, a mile deep. The mineral-rich hydrothermal vents were found, and had the appearance of underwater volcano's at temperatures exceeding 600 degrees, including electrical combustion naturally in volcano's. Life was already thriving through chemical energy from the vents proving that life could thrive without sunlight with extreme temperatures fueled by chemical energy, combustion, and pressure from the depths of the ocean. In a deep-sea vent, all the ingredients of life were already there with all the necessary chemicals of land minerals of life.

Ocean floor volcanic hydrothermal vents

Volcano hydrothermal chemical vents

Ocean floor volcanic hydrothermal chemical vents

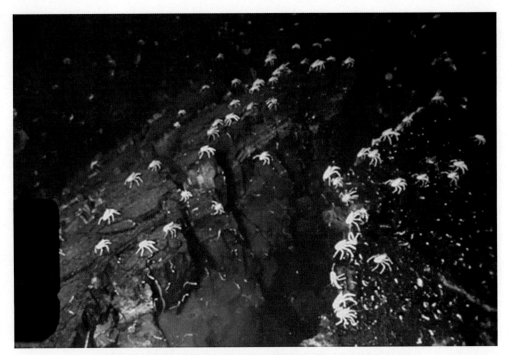

Thriving life from ocean floor thermal vents

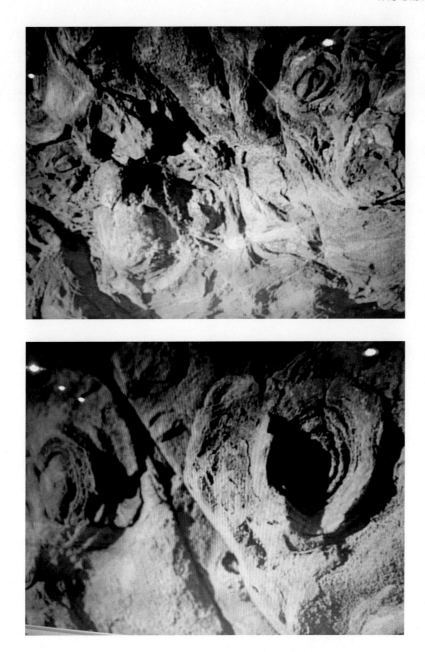

Stromatolite Fossils 3.5 billion years old

Minerals from rocks form organic molecules of life, and these minerals provide cells, and cell surfaces where necessary chemical reactions of life take place. Fossils of earth's earliest life were found in rocks and were formed 3.5 billion years ago. These fossilized rocks were called Stromatolites, and were previously the seafloor, approximately 3.4 billion years ago.

They are the oldest fossils on earth. Stromatolites were found as early fossilized microbial remnants of life. However, they are still living microbes (bacteria) constructing self-made large round mounds underwater, but they are rare. They are similar to the coral mineral structures built in layers. The Stromatolites capture minerals, and sand in the water, then cement them with their biological enzyme activity, layer by layer into large coral-like mounds. The same microbial bacteria use there enzymes to convert the sun's energy into oxygen. They also use their enzymes during the fermentation process, such as winemaking from grapes. They are spectacular wonders of life, and a necessary part of our living process, including eliminating our internal intestinal waste. There wouldn't be life without them.

Stromatolites building layers of minerals

Stromatolite Mounds

Minerals provide cells, and cell surfaces where chemical reactions of life take place. Stromatolites are the oldest direct evidence of life on earth, beginning 3.5 billion years ago as microbial life. Life could have been started earlier than the Stromatolites because of bacterial life to create the Stromatolites, but there is no evidence to be found because bacteria do not leave fossils; however, they do leave chemical signatures.

When bacteria consume nutrients, they produce energy, and carry the chemical footprint of life. Humans also leave chemical footprints of life. Humans take in oxygen and release carbon dioxide. The oxygen reacted with dissolved iron, which then rusted in the ocean, and sank to the bottom of the ocean. The microbes creating oxygen were the cause of miles, and trillions of tons of rust in the ocean waters of iron ore. During this bacterial process, new types of minerals were developed. The estimate from meteorites supplying minerals to earth was approximately two hundred and fifty. The new minerals created from the microbial oxygen created nearly five thousand minerals.

Earth then turned from red to blue from iron ore rust, and then to white, and went into a frozen state again. Active volcanoes fortunately erupted, and broke through the frozen ice that produces large amounts of carbon dioxide, which heats up the planet, thus producing the 'Green House Effect'. The planet got extremely hot and melted, which saved earth from a catastrophic failure of the production of life.

Cliffs of ocean iron ore rust previously underwater

These previously underwater cliffs with iron ore turned the planet red as they surfaced to land. As the ocean cliffs surfaced to land during evolutionary changes, they kept the red iron ore color from the rust at the bottom of the ocean as land and ocean broke apart. The following variations of the earth are next.

Earth turned red from iron and oxygen mixture

Earth froze again then melted and turned blue

Allan R. Rudison, Ph.D.

Volcanic heat broke through the frozen planet

Volcano began to melt the frozen earth

Volcanic heat began to cool into lava lake rock

Lava flowing down the frozen earth after eruption

Volcano's carbon dioxide rises into the atmosphere

Volcanic Lava Lake saturated with minerals of life

Lava Lake hardening into solid rock

The blanket of carbon dioxide warmed the earth

The planet melted because of the blanket of carbon dioxide creating heat and saved the conditions for life to continue. Life expanded in the presence of newly formed oxygen production from the trillions of microbes from the ocean through photosynthesis from the sun's energy. The conversion of sunlight to energy by microbes through photosynthesis is the most significant biological phenomenon of life. No chemical reaction of life comes close to that chemical formula of life on earth from microbes producing vast amounts of oxygen, and additional new minerals from the oceans totaling approximately five thousand.

The rise of minerals could only happen with, and in the presence of oxygen and life. The chemical formula of life on earth from microbes producing vast amounts of oxygen, and additional new minerals from the oceans totaling approximately five thousand. As the earth progressed to change, it began to break apart, creating different climate changes. As a consequence of the earth changes, more oxygen was created and was the cause for larger animals like buffalo and giant sharks.

Water returned to the earth after melting

Earth became habitable 540 million years ago

Early earth oxygen increase caused larger creatures

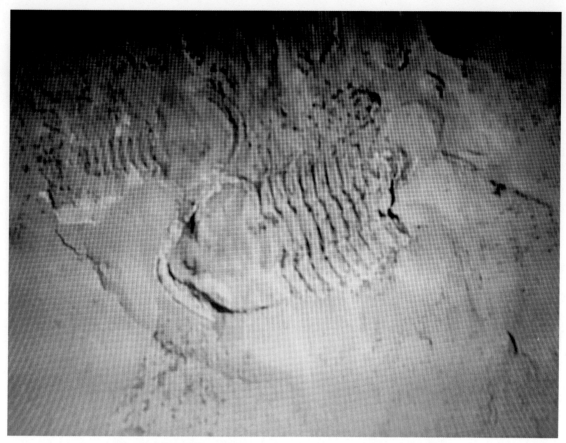

The oldest fossil found on earth is the Trilobite

Rocks also are reasons why life forms developed protective shields to sustain a life of defense. The trilobite developed a calcium carbonate shield from the minerals from rocks in the ocean. The introduction of oxygen was the cause of the massive increase in minerals. Rocks are essential in the creation of life through minerals, and life is essential in creating protective rock shields as in the trilobite development. They developed shells. Stromatolites are the oldest direct evidence of life on earth, beginning 3.5 billion years ago as microbial life. As stated earlier, life could have been started earlier than the Stromatolites because of bacterial life to create the Stromatolites, but there is no evidence to be found because bacteria do not leave fossils.

Allan R. Rudison, Ph.D.

Trilobites on the ocean floor with protective shells

Trilobite Fossil 3.5 billion years old

The trilobite shells are not only made of limestone, which is calcium carbonate, but they also include other minerals. Limestone was also found in the building of Egyptian Pyramids. By developing these evolutionary shells, they were nearly impossible to attack and destroy, which made them the oldest living fossil on earth and helped today's scientists determine when the earth began, 4.5 billion years ago.

The rocks, minerals, and humans evolved together in life's evolutionary process and of new life forms on land and sea. They are inseparable through billions of years of the history of the earth and have transformed life, and life has transformed the planet with its minerals from rocks, oxygen from microbes, and water from the stages that earth has gone through. The building of cities has changed to the composition of the atmosphere, and oceans in our evolution. The next phase of the problems of the atmosphere is beginning to unfold from Global Warming, AKA, Climate Change.

Rocks, minerals & shells made of calcium carbonate

Earth changes from beginning 4.5 billion years ago

These earth photo illustrations are intended to clarify what color changes the earth went through from the beginning of the earth ancient evolution from **Black with fire** to **Grey with granite** to **Blue with water** to **Red with iron ore rust** to **white with freezing,** and to the **Blue** and **Green** planet we live on today with green earth and blue oceans.

The microbes on Stromatolites through photosynthesis, which led to the gas called oxygen that had never been on earth, was one of the ultimate causes of the beginning of life on the ocean floors. Also, the electrical kick-starter for combustion was the spark to combine the amino acid with other minerals for development necessary for teeth, tissue for organs, and bones with other minerals in the ocean water, which is the solvent and the ultimate necessity for the protein building blocks of life were produced. This process created many new minerals necessary for life's process, and combined with the oxygen from bacteria to create life on the ocean floors through volcanic thermal vents providing chemical energy.

The chemical energy from the vents plus the oxygen provided through photosynthesis by trillions of microbes, and the electrical impulse was enough energy for the beginning of life, which initially developed in underwater volcanic oceans makes the best biological sense. We are on the earth; the earth is in the Universe; therefore, 'the Universe in Us', and that's why the evolution of the Dark Matter, Dark Energy expansion of the Universe is directly proportional to the expansion of our brains.

This is why both are accelerating proportionately, and for humans, we see it in our computers, and mobile phones as devices to help us keep up with the daily life's demands of our brains to function by increasing neural growth. The innovation of these devices increases brain neural growth in bytes, such as megabytes, gigabytes, and terabytes, just as in the computers developed by humans and named as such. The concept of the electrical power supply necessity in the Universe for expansion of Dark Matter to Dark Energy is also relative for its expansion of the development for all life. Stanley Miller, from the University of Chicago and Nobel Prize Winner, discovered in the laboratory that the spark of electricity was the kick starter for the production of amino acids mixing with other minerals within a water solvent, and water is a conductor of electricity.

"Science, like nothing else among the institutions of mankind, grows like a weed every year. All other institutions remain the same.

Art is subject to arbitrary fashion (individual perception). **Religion** is inwardly focused and driven only to sustain itself. **Law** shuttles between freeing us, and enslaving us. " - *Kary Mullis*

Science literacy is a vaccination against the world around you! Hopefully, you have been *inoculated* by **"The Universe is in Us"!**

Printed in the United States
by Baker & Taylor Publisher Services